FOREST CREATURES

PORCUPINES IN THE FOREST

SOPHIE WASHBURNE

Gareth Stevens
PUBLISHING

Please visit our website, www.garethstevens.com. For a free color catalog of all our high-quality books, call toll free 1-800-542-2595 or fax 1-877-542-2596.

Library of Congress Cataloging-in-Publication Data

Names: Washburne, Sophie, author.
Title: Porcupines in the forest / Sophie Washburne.
Description: New York, NY : Gareth Stevens Publishing, [2023] | Series: Forest creatures | Includes index.
Identifiers: LCCN 2022005441 | ISBN 9781538279212 (library binding) | ISBN 9781538279199 (paperback) | ISBN 9781538279205 (set) | ISBN 9781538279229 (ebook)
Subjects: LCSH: Porcupines—Juvenile literature. | Forest animals—Juvenile literature.
Classification: LCC QL737.R652 W37 2023 | DDC 599.35/97—dc23/eng/20220214
LC record available at https://lccn.loc.gov/2022005441

Published in 2023 by
Gareth Stevens Publishing
29 East 21st Street
New York, NY 10010

Portions of this work were originally authored by Sebastian Avery and published as *Porcupines*. All new material in this edition was authored by Sophie Washburne.

Editor: Jennifer Lombardo
Designer: Rachel Rising

Photo credits: cover 5213P/Shutterstock.com; cover, pp. 1, 3, 4, 6–8, 10, 12, 14, 16, 18, 20, 22–24 lavendertime/Shutterstock .com; cover, p. 1 furtseff/Shutterstock.com; cover, pp. 1, 3, 4, 6, 8, 10, 12, 14, 16, 18, 20, 22–24 Zvezdesign/Shutterstock.com; p. 5 RudiErnst/Shutterstock.com; p. 7 moosehenderson/Shutterstock.com; p. 7 R. Maximiliane/Shutterstock.com; p. 9 Jens Otte/ Shutterstock.com; p. 11 Warren Metcalf/Shutterstock.com; p. 11 supakrit tirayasupasin/Shutterstock.com; p. 13 Jim Cumming/ Shutterstock.com; p. 15 dangdumrong/Shutterstock.com; p. 17 Custom Life Science Images/Alamy Stock Photo; p. 17 meunierd/Shutterstock.com; p. 19 slowmotiongli/Shutterstock.com; p. 21 Design Pics Inc/Alamy Stock Photo.

Printed in the United States of America

CPSIA compliance information: Batch #CSGS23: For further information, contact Gareth Stevens, New York, New York, at 1-800-542-2595.

CONTENTS

Words in the glossary appear in **bold** type the first time they are used in the text.

A SHARP ANIMAL

Porcupines are large **rodents** that are covered in quills. These quills are sharp like needles, and they help the porcupine **protect** itself from predators.

There are more than 20 species, or kinds, of porcupines in the world. The ones that live in Europe, Asia, and Africa are called Old World porcupines. The ones that live in North, Central, and South America are New World porcupines. The only one that lives in the United States is the North American porcupine.

FIND OUT MORE!

The name "porcupine" comes from the Latin words *porcus*, which means "pig," and *spina*, which means "thorn." In other words, porcupine means "thorn pig"!

Many Old World porcupines look like they have wild hairdos.

OLD AND NEW

Old World porcupines spend all their time on the ground. Their quills grow down the middle of their head and around their back in a **crest**. These quills can grow as long as 20 inches (51 cm)!

New World porcupines spend most of their time in trees. Their quills grow up to only 4 inches (10 cm), but they have little hooks called barbs on them. This makes the quills get stuck in the skin of anything that tries to attack the porcupine.

FIND OUT MORE!

Long-tailed porcupines and African crested porcupines are two species of Old World porcupines. Mexican hairy dwarf porcupines and stump-tailed porcupines are two New World species.

OLD WORLD

- quills don't have barbs
- often grow very large
- quills can grow up to 20 inches (51 cm)
- live in Europe, Asia, and Africa
- fur is very spotted
- can't climb tress

There are many differences between Old World and New World porcupines.

NEW WORLD

- quills do have barbs
- smaller than Old World species
- quills can grow up to 4 inches (10 cm)
- live in North and South America
- can climb trees

LIVING IN THE FOREST

The forest is the perfect **habitat** for the North American porcupine. This species uses trees for both food and shelter, and these porcupines are great at climbing. They often make homes, called dens, inside hollow trees.

Some North American porcupines live where there aren't many trees, such as the desert and even the **tundra**. They are very adaptable, which means they can change to fit their surroundings. They can survive in both warm and cold places.

FIND OUT MORE!

The North American porcupine is the only tree porcupine species without a **prehensile** tail. This makes it hard to balance, so these porcupines often fall out of trees.

Look up in the forest and you might spot a porcupine.

HOW TO SPOT A PORCUPINE

Many New World porcupines are small, but the North American porcupine is one of the largest porcupine species in the world. North American porcupines can grow up to 3 feet (1 m) long with a tail that is 10 inches (25 cm) long. They can weigh up to 35 pounds (16 kg). Each porcupine may have more than 30,000 quills covering its body. Between its quills, it has black or brown fur.

Porcupines have long front teeth, like most rodents. They use these teeth to chew on wood.

FIND OUT MORE!

Hedgehogs are another spiky creature, but they're not related to porcupines at all. They aren't rodents, and they aren't **native** to North or South America.

HEDGEHOGS

One way to tell porcupines apart from hedgehogs is that porcupines are much bigger.

PORCUPINES

STRONG TEETH

A porcupine's large teeth help it chew on all sorts of wood. You can tell which trees porcupines have eaten from because the bark is stripped from the trunk. This sometimes kills trees, so some people think porcupines are **pests**.

Porcupines also eat other plant parts, such as flower buds, fruit, stems, and seeds. They use their sharp teeth to crack nuts. They love to eat the small buds and twigs at the ends of tree branches.

FIND OUT MORE!

When a porcupine falls off the edge of a branch, it sometimes pokes itself with its quills. However, its skin has **antibiotics** in it to help it heal from these cuts!

This porcupine is showing its teeth.

PORCUPINE FAMILIES

Porcupines are solitary, which means they like to live alone. However, males and females come together to **mate** each year.

Mother porcupines usually give birth to only one baby at a time. The baby can't take care of itself right away because its quills are soft when it's born. The mother feeds it and keeps it safe until it's around two months old. When it's between six months and one year old, the baby leaves its mother.

FIND OUT MORE!

Male porcupines are called boars, females are called sows, and babies are called porcupettes. A group of porcupines is called a prickle!

Porcupettes are too small and soft to protect themselves.

A CLOSER LOOK AT QUILLS

Porcupines would much rather scare a predator off than fight. To do this, they raise their quills when they feel **threatened**. If a predator touches or bites the porcupine, the sharp quills stab the predator and get stuck in its skin. The porcupine grows new quills later.

Quills are hollow and made of keratin, which is the same stuff our hair and fingernails are made from. Some people think porcupines can shoot their quills like arrows, but this isn't true.

FIND OUT MORE!

North American porcupines are great swimmers. Their hollow quills make them light enough to easily stay afloat, the way a bird's hollow bones help it fly.

This photo shows the barbs on a porcupine's quill. Under a microscope, the barbs look a little like scales.

STAYING SAFE

Porcupines need their quills to protect them from predators such as fishers. A fisher is a small animal in the weasel family. It knows how to flip a porcupine on its back and attack its belly, where there aren't any quills.

In some places, if a porcupine population gets so large that it's unhealthy for the animals, scientists bring fishers in to lower the porcupine population. Porcupines are also targeted by birds of prey, such as eagles and great horned owls.

FIND OUT MORE!

Old World porcupines' main predators are lions, hyenas, and large snakes called pythons. They're also hunted by humans who want to eat them or use their quills for decoration.

The fisher is one of the biggest threats to the North American porcupine.

A PET PORCUPINE?

Most people never see a porcupine in the wild because porcupines usually come out only at night. During the day, when people are awake, porcupines would rather stay in their den or hide up in a tree. If you do see a porcupine, don't be scared. They don't attack people.

Some states let people keep porcupines as pets. People say they can be friendly once they get used to people. Some of them even like to be petted!

FIND OUT MORE!

Most people don't keep pet porcupines because they need a lot of space, they aren't very cuddly, and they smell worse than dogs or cats.

Pet owners have to be very careful when picking up their porcupine.

GLOSSARY

antibiotic: A drug that can kill germs, including harmful bacteria.

crest: A showy growth on the head of an animal.

habitat: The natural place where an animal or plant lives.

mate: To come together to make babies.

native: Living or growing naturally in a particular place.

pest: A creature that causes problems for people.

prehensile: Able to grasp and hold objects.

protect: To keep safe.

rodent: A small, furry animal with large front teeth, such as a mouse or rat.

threatened: Likely to suffer harm.

tundra: Cold northern lands that lack forests and have permanently frozen soil below the surface.

FOR MORE INFORMATION

BOOKS

Colby, Jennifer. *Porcupine Quills to Needles*. Ann Arbor, MI: Cherry Lake Publishing, 2019.

Leaf, Christina. *Hedgehog or Porcupine?* Minneapolis, MN: Bellwether Media, 2021.

WEBSITES

Fun Porcupine Facts for Kids
easyscienceforkids.com/all-about-porcupines/
Check out more about porcupines here.

National Geographic Kids: Porcupine
kids.nationalgeographic.com/animals/mammals/facts/porcupine
Read more interesting facts about porcupines.

INDEX